PROJECT-BASED LEARNING IN SCIENCE™

BUILDING YOUR PROJECT PORTFOLIO

WITH SCIENCE RESEARCH PROJECTS

CATHLEEN SMALL

rosen publishing's rosen central®
New York

Published in 2019 by The Rosen Publishing Group, Inc.
29 East 21st Street, New York, NY 10010

First Edition

Library of Congress Cataloging-in-Publication Data

Names: Small, Cathleen, author.
Title: Building your project portfolio with science research projects / Cathleen Small.
Description: New York : Rosen Central, 2019. | Series: Project-based learning in science | Audience: Grades 5–8. | Includes bibliographical references and index.
Identifiers: LCCN 2018015263| ISBN 9781508184423 (library bound) | ISBN 9781508184416 (pbk.)
Subjects: LCSH: Science—Study and teaching (Elementary)—Activity programs. | Science—Study and teaching (Middle school)—Activity programs.
Classification: LCC Q182.3 .S5854 2019 | DDC 507.8—dc23
LC record available at https://lccn.loc.gov/2018015263

Manufactured in the United States of America

CONTENTS

INTRODUCTION

What if, instead of taking endless tests to show your scientific knowledge, you could design your own experiments and collect them into a body of work that you could proudly show your teacher and classmates? What if those experiments could be designed around your own interests, rather than being questions generated by someone else? If those things sound appealing, then you will likely enjoy building a project portfolio using project-based learning.

Project portfolios allow students to showcase their best work done over a stretch of time, such as a school year. Project portfolios don't contain everything a student has done—they contain the best of the best. Usually, they include a range of projects, rather than multiple versions of similar types of work.

In science, there are four types of investigations. In field investigations, students observe natural or manmade environments and formulate hypotheses and conclusions based on their observations. One important piece of field investigations is that students must not intrude in the environment—they are strictly

For students interested in the natural world, project portfolios can include botanical investigations such as controlled investigations with plants or field observations of wild or not-so-wild places.

observers. If students interfere in the environment, the investigation generally becomes a controlled investigation.

Controlled investigations involve identifying a problem, collecting some background information, and then generating a hypothesis. At that point, students design a procedure in which they manipulate the variable being studied. As a very simple example, if the hypothesis were that sunflowers grow better in shade than sun, the sun would be that variable. The experiment would involve sunflowers being grown in shade, partial shade, and full sun, always with the same amount of water, the same soil nutrients, and so on. The plant, water, and soil are controlled variables, and the amount of sun is the independent variable.

The third type of investigation is a design investigation. Students identify a need, do background research about the problem, and then generate design ideas that might provide a solution. Once they've settled on the best possible design, they create a prototype and test it out to see whether it did indeed address the problem.

These first three types of investigations are all primary research investigations. Students do background research about the issue or problem in all three, but the actual investigations are all designed and carried out by the students themselves—rather than their having to rely on another scientist's work.

The fourth type of investigation is secondary research. In this type, students identify a problem and do background research to form a hypothesis. They then collect data (which sometimes has been gathered by another scientist, person, or group) and analyze it to try to form a conclusion about their hypothesis. For example, if a student was interested in knowing how the increase in hybrid cars on California roadways has impacted the air quality in that state, they could collect data from the California Air Resources Board or other similar entities. Students would arrange that data and analyze it to form conclusions.

Project-based learning brings scientific investigation to life. In the past, students often studied the works of earlier scientists. And certainly, there is value in learning from the past. But project-based learning allows students to dive right into scientific inquiry, generating their own questions based on their own interests and allowing them to design investigations that lead to thoughtful, real-world solutions.

This book presents a number of projects addressing the four types of investigations, all with project-based learning in mind. The projects will cover physical science, life science, earth and space science, and engineering design. These projects are just a jumping-off point to inspire students to ask their own questions, develop their own hypotheses, and carry out their own investigations.

CHAPTER ONE

PHYSICAL SCIENCE

Physical science encompasses a number of different areas. Loosely, it involves the study of inanimate natural objects. One branch of physical science involves matter and its interactions. This includes the structure and properties of matter, such as atoms and molecules, as well as the chemical reactions that can result from these types of matter.

Another branch of physical science involves ideas of physics, including motion, stability, forces, and interactions. Gravity and magnetic forces are covered in this type of physical science.

Similarly, energy is covered in physical science. This involves temperature, energy transfer, kinetic and stored energy, and how different types of energy can be harnessed and used. It also involves waves, such as sound waves, electromagnetic waves, and ultrasonic waves.

QUESTION: HOW IS THE RATE OF DIFFUSION AFFECTED BY TEMPERATURE OR CONTAINER SHAPE?

Spectators at an event don't clump together in one place. They spread out to fill the space. The same principle holds true for particles in gas or liquid.

Diffusion is the movement of particles from areas of higher concentration to areas of lower concentration. People aren't particles, but if they were, you could think about diffusion in terms of a gymnasium crowded for a popular sporting event. People don't cluster all in one part of the bleachers to watch the big game, squished against each other as close as possible to pack as many into that small space. Instead, the tend to spread out so that everyone has a comfortable place to sit, eventually filling up all the bleachers if the event is popular enough!

The particles in a fluid (gas or liquid) do the same thing. Typically, they don't want to be clustered in small spaces. They want to spread out to fill up a larger space. Diffusion is considered a passive process because no additional energy is required for the particles to move from areas of higher concentration to areas of lower concentration. Diffusion is an example of Brownian motion, which is a physics principle that describes the seemingly random movement of particles in liquid or gas. However, the movement isn't really random—it's caused when vibrating particles crash into each other and move to areas of lower concentration, where they won't be bumping into so many other particles.

All atoms vibrate, even though it's not visible to the naked eye. The speed and force with which they vibrate, however, is partially determined by temperature. The faster the atoms vibrate, the faster they move. However, their rate of movement can also be affected by the shape of container they're in.

PROJECT
COLLIDING PARTICLES

Food coloring is a great way to see diffusion in action. If you put a couple of drops of red food coloring in a glass of water, the drops will start as a concentrated red clump, but they will quickly begin to spread. Before you know it, you'll have a glass of red liquid.

However, this begs the question: what affects the rate of diffusion in water? In this controlled experiment, you'll explore the factors that affect the rate of diffusion.

- *Gather your materials:*
  - *Two clear drinking glasses of the same size and shape*
  - *Two additional clear containers that are a*

different shape from the glasses but hold roughly the same amount of liquid
- *Red and blue food coloring*
- *Hot and cold water*
- *Stopwatch*

- *Fill one glass with hot water. Fill the second one with cold water.*
- *Place two drops of red food coloring in the glass with hot water. Place two drops of blue food coloring in the glass with cold water.*
- *Watch how the color disperses. Takes notes on your observations. Time exactly how long it takes for each color to fully disperse.*
- *Fill the two different-shaped containers with the same amount and same temperature of water. The water temperature should be the same in*

OIL AND WATER

Food coloring is a great way to see diffusion in action, but not all liquids mix uniformly like water and food coloring. Oil and water, for example, don't mix. If you put food coloring in water (to make it easy to distinguish from the color of the oil) and then pour some oil in a glass with the water, you'll find that the oil will separate and float to the top. That's because water is denser than oil (and thus stays at the bottom of the glass), and water molecules are more attracted to each other than they are to oil molecules. So instead of bouncing into each other and mixing uniformly, the water and oil molecules stay apart.

Food coloring is a great way to study diffusion because the vibrant colors are easy to watch spread throughout most liquids.

both containers because it is a controlled variable here; the independent variable is the shape of the container.

- *Place two drops of red food coloring in one container and two drops of blue food coloring in the other.*
- *Observe how the color disperses in both containers. Take notes of your observations. Again, use the stopwatch to gather exact data about the rate of dispersal.*
- *Arrange your data in a presentation of your findings. Include a chart or graph. Share your conclusions about how temperature and the shape of the container individually affect the rate of diffusion in water.*

QUESTION: HOW DOES STEAM CREATE MOTION?

Steam engines have been around for centuries. They were used to pump water out of coal mines, drive electrical power generators, and even power lights. But they are perhaps best known as the force behind the steam locomotive. These engines pulled trains from the early nineteenth century through the mid-twentieth century.

But how could steam, which is simply the gaseous vapor that water is converted into when it's heated, power giant locomotives? Steam vapor disperses in the air quickly—where is the force in that?

As you learned in the previous project, particles are constantly moving, especially when heated. Steam is no exception—there is potential energy in steam particles. The question was how to convert that potential energy into kinetic energy to move vehicles. The

Steam engines use physics properties to harness the energy held in vapor. This steam engine is pulling a train along the Durango & Silverton Narrow Gauge Railroad in Colorado.

answer involves not only Brownian motion, but also Newton's Third Law of Motion: that for every force or action, there is an equal and opposite reaction.

PROJECT
STEAM-POWERED VEHICLES

A boat is a good way to see how steam can move a vehicle, using both Brownian motion and Newton's Third Law. It is easy to build a boat to test this in any body of water—even in a bathtub! This experiment will let you see the physics of steam in motion.

- *Gather your materials:*
  - *Thick Styrofoam board*
  - *Untwisted wire coat hanger*
  - *Box cutter*
  - *Ruler*
  - *Cork*
  - *Stainless-steel pipe with a cap on one end*
  - *Two tea candles*
  - *Two nails of the same size*
  - *Three nails of different sizes*
  - *Hammer*
  - *Modeling clay*
  - *Water, and a way to boil it*
  - *Glue or tape*
  - *Lighter or matches*
  - *Notebook and pencil or pen*
  - *Stopwatch*

- *Wrap the coat-hanger wire firmly around the front and back of the stainless-steel pipe so that it is securely wrapped and has equal lengths of wire sticking out from each side.*
- *Use the ruler to draw a boat shape on the Styrofoam board. The boat should be about 4 inches (10 centimeters) wide and about 7 inches (18 cm) long. It should look like a rectangle, but one short end of the rectangle should be triangular. This will be the front of your boat. If you stood it on the opposite short end, it would look like a rectangular building with a peaked roof.*
- *Have an adult help you with this step, as box cutters can give you a nasty cut. Use the box cutter to cut out the boat shape.*
- *Mount the stainless-steel pipe on the boat by inserting the coat-hanger wire ends into the Styrofoam. The pipe should run lengthwise, from short end to triangle end of the Styrofoam, and should be elevated over the Styrofoam so that the tea candles can later be placed underneath the pipe, on top of the Styrofoam boat. The coat-hanger wire should to be thick enough to easily hold the pipe up above the Styrofoam; if it isn't, look for a thicker hanger. Make sure the capped end of the pipe is at the front, triangular end of the boat.*
- *Seal the holes made by the wire in the Styrofoam with modeling clay. Otherwise, your boat may leak and sink!*

- *Gently hammer the two same-size nails into the front and back of the Styrofoam boat, from top to bottom. They should be equally spaced widthwise, with one nail near the short end of the boat and one near the point of the triangle at the front. These nails will help stabilize the boat, much as a keel or rudder on a real boat does.*
- *Seal the nail holes with modeling clay, again to prevent leaking.*
- *Glue or tape the tea candles to the top of the Styrofoam boat, beneath the stainless-steel pipe. One should be near the front of the pipe and one near the back.*
- *Use the smallest of the remaining nails and hammer it into the center of the cork to create a small hole. (Remove the nail after you've made the hole.)*
- *Now it's time to prepare the body of water in which you can test your boat. You might fill up the bathtub or a large basin partway. Or you could use a swimming pool or pond, but be sure you can retrieve your boat after the experiment!*
- *Going back to the boat, have an adult help you with the next step. Boil some water and let it cool slightly. You can boil it in a kettle or in a microwave.*
- *Pour approximately half a cup of hot water into the open end of the pipe. Be careful not to burn yourself with the hot water, and*

remember that the steel pipe will heat up with the water.

- Place the cork into the open end of the pipe.
- Light the candles beneath the pipe. Again, be careful not to burn yourself!
- Place the boat in the body of water you've prepared or selected and watch what happens. Record your observations. Use the stopwatch and record how long it takes the boat to move to a certain designated location.
- Retrieve the boat, blow out both of the candles, remove the cork, and pour the water out of the pipe. Careful! It will be hot.
- Using the hammer and the next larger nail, make the hole in the cork slightly larger.
- Boil more water, pour approximately half a cup in the pipe, and recork the pipe.
- Light the candles again and place the boat in the body of water at the same place you did before.

Steam will escape through the hole in the cork in the same way it escapes through the hole in a teakettle spout.

- *Watch what happens and record your observations again. Time how long it takes the boat to reach the same designated location as before. Was it faster or slower?*
- *Retrieve the boat, blow out the candles, remove the cork, and pour out the water.*
- *Using the hammer and the largest nail, make the hole in the cork slightly larger.*
- *Boil more water, pour approximately half a cup in the pipe, and recork the pipe.*
- *Light the candles again and place the boat in the body of water at the same place you did before.*
- *Watch what happens and record your observations again. Make sure to use the stopwatch to time how long it takes the boat to reach the designated location.*
- *Arrange your data in a presentation of your findings. This might be in a chart or graph, for example. Share your conclusions about how the steam propels the boat and how increasing the size of the hole in the cork affects the boat's power and speed.*

CHAPTER TWO

LIFE SCIENCE

Life sciences deal with living things—as the name suggests! All living organisms are composed of cells, so life sciences start at the cellular level and work their way up through whole organisms—as simple as single-celled organisms and as complex as the human body.

All living organisms have genetic traits that are passed from one generation to the next. This is true for very simple organisms, as well as complex organisms such as humans. How that genetic information is passed to future generations depends on the organism and its method of reproduction. Life sciences explore all of these methods, as well as how genetic traits are inherited and how genetic mutations lead to the evolution of species.

Finally, life sciences look at the ecosystems that support organisms. They study organisms' effects on their ecosystems and vice versa. Essentially, they look at the circle of life: how each organism and ecosystem contributes to biodiversity and the ongoing cycle of life on Earth.

QUESTION: HOW DOES THE PLANTING OF DROUGHT-TOLERANT PLANTS HELP CONSERVE WATER?

Years ago, when you drove through suburban neighborhoods, nearly every house had a beautiful green lawn, lush and inviting. It wasn't uncommon to see sprinklers running during the heat of summer days, keeping the lawn verdant and appealing.

In recent years, though, there has been a move toward drought-tolerant landscaping, especially in states that don't get much

Particularly in states with desert environments, such as Arizona, drought-tolerant landscaping using plants like succulents is a popular choice that helps conserve valuable water resources.

natural rainfall, such as California, Nevada, Arizona, and New Mexico. That move has expanded to other states as scientists have shared that climate change and global warming are expected to result in warmer temperatures and drier conditions in states that haven't historically suffered from droughts.

It's said that a global movement is needed to combat climate change, but a global movement always starts with one person. Add up the efforts of enough people, and you get global change. So, how much water can one person save by switching from traditional grass to drought-tolerant grass?

SUCCULENTS: A DROUGHT'S BEST FRIEND

Many succulent plants are native to dry desert locations. For example, cactuses are succulents, and they are found in desert areas such as much of Arizona. (Note, though, that not all succulents are cactuses!) There are about sixty different plant families that contain succulents. They come in almost every color you can think of. Some succulents produce colorful blossoms, and the desert in bloom offers more color than you might imagine. In addition to being drought resistant, succulents tend to be resistant to pests, such as aphids or other insects that attack some plants.

PROJECT

A TALE OF TWO LAWNS

One way to conserve water in landscaping is to go with all drought-tolerant plants, such as

Grass seeds can be planted in small trays and used as the basis for experiments on different varieties of grasses.

native succulents. However, many people like the look of a lawn but still want to conserve water. In this controlled experiment, you will examine how drought-tolerant versus traditional grasses grow when light, soil, and water conditions are kept constant.

- Gather your materials:
    - Three to five grass seed trays
    - Grass seeds: try to get at least one traditional type and several drought-tolerant varieties
    - Enough soil for your trays
    - Grow light (optional)
    - Water
    - Notebook and pencil or pen
- Fill each seed tray with a comparable amount of soil.
- Plant one type of seed in each tray. Be sure to mark your trays with which seeds you used in each one so you can accurately track your results!

- *Place the trays in identical light conditions. Grass typically grows in sunlight, but you could also use a grow light.*
- *Water each tray the same amount each day. Grass typically requires a fair amount of water in the beginning, but be careful not to give the seeds too much water. You'll want to keep the soil moist without overwatering.*
- *When the grasses have all sprouted and are beginning to grow, decrease the amount of water you give them daily. You should still give them an equal amount of water daily, but that amount should be less than before.*
- *Record your observations daily about the growth of each tray and the amount of water you are giving each tray. If all the trays are still growing after you decrease the amount of daily water, decrease it even further and continue to record your observations. You'll want to continue decreasing the amount of water until at least one of the trays stops growing and dies, since the idea is to see which grasses can survive with the least amount of water.*
- *Arrange your data in a presentation of your findings. This might be in a chart or graph, for example. Share your conclusions about which grasses continued to thrive even when the daily water was cut back and how your results can be applied to home landscaping with an eye on water conservation.*

QUESTION: WHAT IS THE PURPOSE OF INSECTS IN A GARDEN ECOSYSTEM?

Ecosystems are an important part of life sciences. Without ecosystems, organisms wouldn't survive—and without organisms, ecosystems wouldn't survive. Birds, bees, and bugs all play vital roles in a garden ecosystem. As much as insects might seem like pests, most of them have important roles in supporting a garden ecosystem.

PROJECT
GARDEN LIFE

In this field investigation, you'll study a specific garden ecosystem and the insects that inhabit it, and then you'll draw conclusions about the role each one plays in the garden ecosystem.

- *Gather your materials:*
  - *Books or other sources about plants and insects native to your area*
  - *Notebook and pencil or pen*
  - *Magnifying glass*
  - *Gloves (optional)*
  - *Camera (optional)*
- *Do some background research on what plants and insects are native to your area. Many gardens contain both native and nonnative plants and insects. Your investigation will be enriched if you know which parts of the*

ecosystem are native to your area and which have been introduced.

- Select a garden. If you have a backyard garden, that's a great place to do your field investigation. Alternatively, you could study a school garden, a community garden, or part of a park.
- Visit the garden daily, at a variety of times. You may see different activity at different times of day, so it's important to gather a good sampling of data.
- On each visit, start by recording the date and time in your notebook.
- Take notes about everything you observe. For insects, you'll want to describe their size, shape, colors, markings, and any sounds they make. Make sketches of what you see in your notebook. That will help you if you need to look them up to identify them later. If you have a camera, you can take pictures of them, too.

By visiting the garden at different times of the day, you may get the chance to see a variety of different insects in the ecosystem.

- *Make notes about the plants in the ecosystem, too. If you have a camera, you can take pictures of them. If you're good at sketching, draw some quick sketches. You can use these later to look up the plants and determine what they are. If you have trouble identifying plants through internet or book research, try visiting a local nursery or nature center. The professionals there are often very good at knowing what plants grow in the area.*
- *Wearing gloves if you want, gently move soil, plants, rocks, and sticks to look for other signs of life. Take care not to disturb the ecosystem—you don't want to damage plants or hurt any insects! But if you just gently move items and insects in the ecosystem, you may find new and interesting things to observe and record in your notebook. Insects in particular sometimes hide under the top layer of soil, beneath decomposing leaves or logs, or in dark spaces under rocks or tree roots.*
- *Watch what the insects do in the garden and record what you see. This will help you draw conclusions about the insects' roles in the garden ecosystem.*
- *With the data you've gathered, create a presentation of your findings. You can detail both the types of plants and the types of insects you saw, as well as what the insects' role in the ecosystem appeared to be.*

QUESTION: WHAT ARE THE MOST EFFECTIVE MATERIALS FOR WORM COMPOSTING?

As people become more environmentally conscious and interested in ways they can recycle and help the environment, worm composting is becoming more common. Contrary to how the name might sound, worm composting doesn't involve composting the worms themselves—that would be cruel! Rather, worm composting refers to the process of using worms to recycle food scraps and organic material into a soil amendment called vermicompost.

How exactly does this work? Quite simply, worms eat and then they poop. Their poop is the vermicompost that helps plants grow. It's estimated that one tablespoon of pure worm vermicompost provides enough nutrients to feed a 6-inch (15 cm) potted plant for at least two months.

PROJECT
FEED THE WORMS, GATHER THE POOP!

To determine the most effective materials for maximizing vermicompost, you have to build some worm bins, get some worms, and start feeding them!

- *Gather your materials:*
  - *A few shallow boxes made of plastic, wood, or glass—about five gallons each is a good size*
  - *Newspapers or leaves*
  - *Red worms (Eisenia foetida) or red wigglers (Lumbricus rubellus)—if your local garden*

shop doesn't have these species, you can order them online
- *A variety of fruit scraps (avoid citrus—it's too acidic)*
- *A variety of vegetable scraps (avoid onions and broccoli—their odor is too strong)*
- *Scale that is capable of weighing small amounts*
- *Measuring spoons and cups*
- *Notebook and pencil*

Certain species of worms are especially efficient at producing rich vermicompost. This worm is a *Lumbricus rubellus*.

- *Prepare your containers. Rinse them out. If you're using wooden boxes, line the bottom with plastic, such as from a plastic bag or shower curtain.*
- *Find a loose-fitting lid for each container. Don't use anything that fits too tightly—the worms need to be able to get some air!*
- *Tear the newspaper into strips and moisten the strips with water. Place them in the containers as "bedding" for the worms. Or you can use leaves—make sure to moisten them slightly.*

- *Place a roughly equal number and type of worms in each bin.*
- *Add food scraps to each bin. Bury the scraps in the bedding. Be sure to label each bin with the type of food scraps you've added, and do not add more than one type of food scrap to each bin.*
- *Cover the bins loosely and put them in a warm (but not hot!), dark place. Worms need warmth, darkness, air, food, and moisture.*
- *Add more food scraps as needed. Be sure the bins stay moist, too. If they seem dry, you can use a spray bottle with a little water to add some moisture.*
- *Observe the bins at regular intervals and note your observations in your notebook. You'll want to check them regularly to add food scraps anyway, so that's a good time to do your observations.*
- *Composting doesn't happen quickly, so this experiment works best if you have at least a*

Creating vermicompost isn't a fast process, but with enough time and the proper food scraps and bedding, worms can eventually produce an impreesive supply of vermicompost.

month to devote to it—preferably longer. After three to five months, the worms will likely have replaced much of the bedding with compost, and you can then harvest the compost and determine which bin had the greatest yield of vermicompost.

- *To harvest, do not add any new food scraps to the bin for two weeks. After two weeks, push all of the contents of the bin to one half of the bin and remove any large pieces of food or newspaper bedding. Place fresh bedding and food in the now-empty side of the bin and continue burying food scraps only in that half. When you replace the cover, cover only the new half. This will help the old half of the bin dry out.*
- *Wait two to three weeks while the worms move to the new side of the bin. They will automatically follow the food and moisture in the covered half of the bin. In doing so, they will leave behind the drying compost on the old side. You can then remove the compost.*
- *Weigh the compost from each bin on the scale and record the weights.*
- *Measure the amount of compost using measuring spoons and cups (depending on how much you have). Record the amounts.*
- *With the data you've gathered, create a presentation of your findings and conclusions about which types of food are most effective for vermicomposting.*

CHAPTER THREE

EARTH AND SPACE SCIENCES

As you probably imagine, earth and space sciences cover Earth and space! In the most general sense, they cover Earth's place in the solar system, the Milky Way galaxy, and the universe. However, they dig much deeper than that. Earth and space sciences cover Earth's motion, the seasons, tides, eclipses, and weather. They look at Earth's crust and inner core. They examine the tectonic plates that make up Earth's crust and how their movement affects the planet. They look at how changes in Earth's systems affect organisms and vice versa. Finally, they examine natural hazards, natural resources (including use and depletion), human impact on natural resources, and global climate change.

Earth and space sciences also involve the study of other planets and objects in our solar system, although in general much less is known about them than is known about Earth. There is still a lot of unknown in the solar system and universe—much more to be explored and studied! Earth has been studied in much greater depth, but because it is constantly changing as a result of natural changes and human intervention, there is much more to be studied.

QUESTION: HOW CAN YOU MAKE USE OF WIND POWER IN YOUR COMMUNITY?

Wind power is becoming an increasingly used alternative energy source. Wind farms are springing up in windy areas where the conditions lend themselves to harnessing the greatest amount of wind energy to be converted into power. Wind farms consist of turbines that turn in the wind, and the wind energy they capture is used as a power source.

Some wind farms are offshore, out in the ocean. But most are on land, and they are found all across the globe. However,

This wind farm is located in Idaho. The mountains in the far background likely produce a decent amount of wind to sweep across the plains and rotate the turbines, creating energy.

six of the ten largest wind farms in the world happen to be in the United States. Three of the top ten are in Texas, one is in the Central Valley of California, one is in Eastern Oregon, and one is in Indiana.

What do all these US locations have in common? They tend to be largely flat areas, free of large mountains that can block or slow the wind. However, mountaintops are notoriously windy, so certainly there are some opportunities for wind energy there as well.

PROJECT
WHERE SHOULD I PUT THAT TURBINE?

How might you determine where in your community would be the best place for a turbine to harness wind energy? This field investigation will help you decide.

You might be able to look up wind velocity data for your community, but that would just give you a general overview. You want to know where specifically in your community a turbine might work well. To do that, you could use the Griggs-Putnam Index of Deformity, which looks at the shape of trees and hedges to determine areas where wind is frequent and/or strong. Typical trees in areas without much wind grow mostly straight up and down. The Griggs-Putnam Index helps scientists look at factors like flagging (when the tree is leaning to one side), brushing (when the small limbs of the tree are bending downward), and throwing (when the tree trunk is bent to one side), and helps them assign a value that corresponds to an estimated wind speed.

- *Gather your materials:*
  - *Notebook and pencil or pen*
  - *Griggs-Putnam Index of Deformity chart*
  - *Camera (optional)*
- *Start by finding and printing out a copy of the Griggs-Putnam Index of Deformity. Simply search the internet for the name of the index, and you'll find numerous sources.*
- *Identify several locations in your community that have different types of terrain, that have trees, and that experience wind at least sometimes. If possible, these should include:*
  - *A flat, open area, such as an athletic field in a park*
  - *Each side of a hill, plus the crest of the hill*
  - *Spots on each side of a large building.*
- *Identify several trees at each of these locations. Three would be ideal, but if that's not possible, you should find at least two trees.*
- *In your notebook, use a different page for each location. Write the name of the location at the top of the page so you don't get your data confused. Write down the trees you're using for data collection in each location.*
- *For each tree, look at the Griggs-Putnam Index of Deformity and assign a rating*

number. The ratings go from 0 to VII (Roman numerals), and each corresponds to an estimated wind speed.

- *For each tree, look for evidence of turbulent winds, such as broken branches around the base of the tree or a significantly leaning trunk. Make note of any evidence of turbulent winds.*
- *If you have a camera, take pictures of each tree.*
- *With the data you've gathered, create a presentation outlining the winds at each location. Draw conclusions about the best place to put a wind turbine.*

The bent nature of this tree gives an idea of the wind direction and the fact that the wind here is likely forceful.

QUESTION: HOW DOES POPULATION AFFECT AIR QUALITY?

In the 1970s, the air quality in Los Angeles was notoriously bad. More than two hundred days each year, the air quality was considered to have unhealthy levels of pollution, and the city and surrounding areas were often covered by a thick blanket of brownish-orange haze.

Since then, regulations put in place by the Environmental Protection Agency and various other state and federal agencies have much improved air quality overall. However, there are still places where air quality is often poor.

THE DIRTIEST AIR

The United States doesn't have a monopoly on air pollution. Far from it, in fact. You might be surprised by which cities worldwide have the worst air pollution. In 2017, some of the global cities with the most polluted air included Lagos, Nigeria; Delhi and Mumbai, India; Karachi, Pakistan; Beijing, China; Bangkok, Thailand; Mexico City, Mexico; Buenos Aires, Argentina; and Dhaka, Bangladesh. The rankings for these cities were based on the levels of four major gases associated with air pollution: ozone, methanol, formic acid, and ammonia. Some of these gases are produced due to burning of various resources, but scientists also found one rather unusual possible contributor to the poor air quality in Buenos Aires: the meat-slaughtering industry there. This is evidence that industry sometimes has unexpected consequences.

Each year, the American Lung Association releases a State of the Air report detailing air quality across the United States. In 2017, many of the cities with the worst air quality were in California, though several were in Texas and a handful of other states as well. You might assume that makes sense, since California is a heavily populated state. More people equals more cars equals more pollution, right? As it turns out, some of the most polluted cities (in California or otherwise) weren't the dense urban metropolises that jump to mind.

PROJECT
BIG CITY, DIRTY AIR?

How does population affect air quality? This experiment will help you draw some conclusions.

- *Create a research question. You're looking at how population appears to affect air quality, but there's more to it than that. You may find heavily populated areas with better air quality than less populated areas, and you'll need to look at what else might cause a particular area to have poor air quality. You can think about geographical features, industry, and transportation trends, for example.*
- *Develop a hypothesis. What do you think your investigation might show about air quality in cities of greater and lesser population?*
- *Select a number of cities or regions across the United States for which you want to study the air quality. Choose both a heavily populated city and a less populated city in each of the following regions: the West, East, South, Northwest, Northeast, and Midwest.*
- *Find and gather your data. The American Lung Association's State of the Air site (http://www.lung.org/our-initiatives/healthy-air/sota)*

The internet makes carrying out secondary research investigations much easier. Information such as air-quality data is very easy to access. It's available at your fingertips!

has a lot of good information. So does the Environmental Protection Agency (https://www.epa.gov/outdoor-air-quality-data).

- *Organize your data and look for trends or similarities. Often a chart or graph will help you easily pick out trends and connections.*
- *Draw conclusions from your data analysis and prepare a presentation that details your findings and conclusions.*

QUESTION: CAN THE FREQUENCY AND MAGNITUDE OF EARTHQUAKES BE PREDICTED?

Pretty much any area in the United States is subject to some type of natural disaster: hurricanes and tropical storms, floods, tornadoes, wildfires, or earthquakes. All of these disasters can be devastating, taking lives and homes. Preparation is often key. Tornado alerts allow people to seek shelter in a basement or other safe area when a tornado is bearing down on their town. Evacuation alerts help people flee areas that are being threatened by wildfires or floods. Hurricanes and tropical storms are often known about days in advance, since they typically form offshore before heading toward land. This gives residents of hurricane-prone areas time to put storm shutters on their homes, gather water and other supplies needed to survive without power or running water, and if necessary leave the area.

Earthquakes are unique in that there is virtually no advance warning. Sometimes small "preshocks" will rattle an area in the days or weeks before a big earthquake, reminding people to take precautions. But often the small earthquakes that people think are preshocks are just minor quakes releasing pressure along the fault lines—in fact, they can help prevent big earthquakes by releasing that pressure!

When a large earthquake hits, there is usually no warning. People simply take cover wherever they are when the ground starts shaking. Those who live in states often rattled by earthquakes, such as California, Washington, and Alaska, live under the constant warning of, "We expect a big quake will strike along this fault sometime in the next few decades."

However, seismologists have been studying earthquakes for years, particularly in the western states where earthquakes are a

In earthquake-prone areas like California, buildings must adhere to stringent codes to ensure safety during a quake. However, even the strongest structures will sometimes collapse when subjected to the force of a quake.

looming threat, and they have collected much data on the various faults that run along the earthquake-prone areas of the United States.

PROJECT
ROCKIN' AND ROLLIN'

Do you think there's any way that data could help predict the frequency of earthquakes? What about the magnitude, or strength, of earthquakes?

- *Create a research question. You're looking at earthquake data, but what exactly do you want to know about it? Are you interested in the frequency of quakes or the magnitude? Are you curious about which faults are the most likely to have a major quake and when that might happen? Think about these issues as you form your research question.*
- *Make a hypothesis. What do you think your investigation might show about earthquakes?*
- *Select a number of faults to study. ScienceStruck has a list of several major faults in the United States (https://sciencestruck.com/fault-lines-in-united-states). If you're interested in California faults, the Southern California Earthquake Data Center (http://scedc.caltech.edu/significant/fault-index.html) has an alphabetized index.*

State and federal websites offer a great amount of earthquake data, particularly about earthquake-prone states such as California, Washington, and Alaska.

- *Find and gather your data. The USGS website (https://earthquake.usgs.gov/data) has a vast database of earthquake data. If you're interested in California earthquake data, be sure to see the Southern California Earthquake Data Center site (http://scedc.caltech.edu).*
- *Organize your data and look for trends or similarities. A chart or graph may help you pick out trends and connections.*
- *Draw conclusions from your data analysis and prepare a presentation that details your findings and conclusions.*

CHAPTER FOUR

ENGINEERING DESIGN

Engineering design is very hands on. This is where you identify and define problems, brainstorm possible solutions, and design products. It can encompass any of the other branches of science as well. For example, if in life science you're studying the effects of tap water on plants and you find that tap water contains elements harmful to plant growth, you might brainstorm possible ideas for a filtration system and build a prototype of a system to filter out harmful elements in the water.

In engineering design, testing is key, too. It's rare that a prototype works perfectly on the first attempt. More often, you test the prototype, identify any design flaws, fix them, and build a second prototype. This cycle repeats until you have the best possible solution engineered, with a working prototype.

Engineering design also involves looking beyond the immediate impact the prototype has. For example, if you built a water-filtration system that successfully filtered out the water elements that were harming your plants, would there be negative impacts on other parts of the environment? Would other plants suffer from the lack of that element? Would worms and insects living in the garden ecosystem be harmed by the lack of the element?

QUESTION: WHAT IS THE BEST WAY TO DESIGN A BRIDGE?

Bridges are one of mankind's great engineering accomplishments. Not only do they have to be built strong enough to carry people, cars, or trains, but they also have to be engineered to be stable in

California's Golden Gate Bridge is a modern marvel of engineering—a suspension bridge that must be strong enough to withstand the strong earthquakes that rattle the region.

unfavorable weather conditions, such as severe wind, hurricanes, earthquakes, and tornadoes.

There are several types of bridges, all with pros and cons in terms of engineering. The simplest and oldest type of bridge is called a beam bridge, or girder bridge. A simple beam bridge has a support at each end with a horizontal structure (the beam) across the top. A good example would be two soup cans with a book laid across the top of them.

A truss bridge is like a beam bridge, except it uses basically triangles as support. Triangles help disperse the compression and tension on the beam so that the beam doesn't buckle at weak points. Sometimes the trusses are above the beam and sometimes they're below, but the result is the same. You can search the internet for pictures of truss bridges to get an idea of what the trusses look like—there are many designs.

Suspension bridges have a beam supported by cables that are run from towers. Two of the best-known suspension bridges in the United States are the Golden Gate Bridge in San Francisco and the George Washington Bridge in New York City.

PROJECT
BUILDING BRIDGES

See for yourself what the advantages of different bridge designs are by building bridge models.

- *Research the different types of bridges to get ideas for your design. Each type of bridge has common features in structure, but there are many small variations to the designs.*
- *Sketch out a blueprint for what you want each of your bridges to look like. Plan to build*

at least one beam bridge, one truss bridge, and one suspension bridge.

- *Decide on the length for each bridge. They should all be roughly the same span.*
- *Gather your materials:*
  - *Popsicle sticks*
  - *Toothpicks*
  - *String*
  - *Hot glue gun and hot glue sticks*

Toothpicks can be used to create models of any of the bridge types. Popsicle sticks are another popular building material for prototypes.

  - Small wooden blocks or some other small item that can be used to test the bridge strength
  - Notebook and pencil or pen
  - Camera (optional)
- Build each bridge. If you want to use materials other than the ones specified, feel free. They are just a starting point for you. Be sure to have a parent help you when you use the hot glue gun—burns from those can be nasty!
- If you'd like, take a picture of each completed bridge before testing it. You may want to use the pictures when you present your findings.
- Add the blocks (or other weights) one at a time to see how much weight your bridge can hold. Be sure to put the weights at roughly the same point on each bridge to be sure you're making an apples-to-apples comparison on your bridges' strength.
- In your notebook, keep notes about the materials you used to build each bridge, as well as how it withstood the weight. How much weight could each bridge hold before it became unstable?
- Form conclusions from your designs about which bridge type was the strongest and what ways you might improve upon the structure of each bridge to make it stronger. Prepare a presentation detailing your findings.

QUESTION: HOW CAN YOU COLLECT WATER IN DRY ENVIRONMENTS?

The United States has many areas where water is plentiful, but it also has many areas where drought and dry conditions are an everyday problem. Deserts are obviously dry, but there are also other regions that have surprisingly little water. For example, California borders an ocean, yet drought is a constant problem because of low yearly rainfall!

People in regions where water can be scarce find creative ways to conserve and collect water. They put buckets under their showerheads

During the 2012–2016 drought, California's Shasta Reservoir dropped to alarmingly low levels. Fortunately, the state experienced a wet winter in 2016–2017 that restored water to more normal levels.

to gather cold water as they wait for their shower to heat up. They then use that cold water to water their gardens. Some buy more efficient water heaters, dishwashers, and washing machines to conserve water.

WEIRD WAYS TO SAVE WATER

In addition to the usual ways to save water, there are some unusual ways you might not have considered! Some people collect the "gray water" from the washing machine after a load of clothes was washed and use that water to mop their floors. Or, they install a filtration system so they can use the gray water to water their lawn. Some people who collect a bucket of cold water while the shower heats up then pour that bucket of water into the toilet to manually flush it the next time they use the toilet. Each toilet flush uses approximately a gallon and a half of water, so some people choose to "let the yellow mellow" in the toilet and not flush, or they urinate in the shower so the urine will wash down the drain. Water conscious? Yes. Sanitary? Not really.

Rainfall-collection tanks are an easy way to gather water that can be used for irrigation, but they're only really useful in areas where it rains a fair amount. However, there's a possible solution in areas where rainfall is scarce: dew traps.

PROJECT
BUILDING DEW TRAPS

Dew forms nearly everywhere, even in dry regions. It is the result of water vapor in the air cooling in the colder hours and forming water droplets. You might see dew on your lawn or outdoor plants in the early mornings during cooler months or on the cars you walk by on the street early in the morning. The trick with dew is that it forms on a surface that it won't soak into. Soil and dirt automatically soak up water, so dew doesn't stay on them for long. But with a little engineering, you can make a dew trap to collect water.

- Gather your materials:
  - Shovel
  - Aluminum pie plates for water collection (You can also use food-storage containers, but make sure they're all the same size and shape.)
  - Two 9 x 12 foot (about 2.7 x 3.7 meter) clear plastic drop cloths, of 2mm thickness (found at most hardware stores)
  - Scissors
  - Ruler
  - Bricks
  - Egg-sized rocks
  - Disposable cups
  - Cheesecloth (found at most grocery stores or online)
  - Rubber band
  - 50ml graduated cylinder (found anywhere that sells science supplies or online)
  - Notebook with pencil or pen

- *Locate a sunny place in the dirt where you can dig multiple holes. Be sure the holes aren't any place where people might accidentally step in them.*
- *Dig the holes. Make sure the holes are relatively flat at the bottom so that your collection pan will lie flat.*
  - The first hole should be 1 foot (30 cm) in diameter at the top and 1 foot (30 cm) deep. The bottom of the hole should

Digging holes can be surprisingly difficult in dry, sun-parched areas. But be persistent. You will need to make reasonably deep holes for this experiment to work.

  - be just wide enough to fit the collection pan. Use the ruler to check the diameter and depth.
  - The second hole should be 2 feet (61 cm) in diameter at the top and 1 foot (30 cm) in depth. Again, the bottom of the hole should be just wide enough to fit the collection pan.
  - Dig at least two more holes of different diameters. As before, the bottom of the hole should be just wide enough to fit the collection pan.
- *Place a collection pan in the bottom of each of the holes you dug.*
- *Use the scissors to cut plastic circles for each hole. For the narrower hole, you'll want a plastic circle that's 4 feet (2.1 m) in diameter. For the wider hole, you'll want a plastic circle that's 5 feet (1.5 m) in diameter. Size up the plastic circles accordingly for your other holes.*
- *Center the plastic circle over the appropriate hole. Put bricks around the edge of each hole to hold the plastic in place.*
- *Gently press down in the center of the plastic for each hole so that the plastic circle forms a cone-shaped "bowl" above each collection pan. This will make the condensation run down and pool in the cone end of the plastic.*
- *Put an egg-sized rock into each plastic "cone." Center it over the collection pan.*

- *Allow the dew trap to collect water for several days. Then, measure the dew collected. Carefully remove the bricks from the edge of the plastic.*
- *Lift the edges of the plastic up together, forming a sort of bag in which the pooled dew is in the bottom. The egg-sized rock will still be in there, too. Be sure to keep holding it over your collection pan in the hole.*
- *Use the scissors to gently slice open the bottom of the plastic "bag" you're holding, so that the collected dew pours out into the collection pan.*
- *When all of the water has drained, remove the collection pan carefully from the hole.*
- *Hopefully, you've avoided getting much dirt in your collection pan. It's hard to keep all dirt out, so now you can strain the liquid to accurately measure how much you collected.*
- *Cut a piece of cheesecloth and fold it in half so that it fits over the mouth of a plastic cup. Secure the cheesecloth to the cup with the rubber band. This will act as your strainer.*
- *Slowly and carefully pour the water from the collection pan into the plastic cup.*
- *Remove the cheesecloth and slowly pour the water from the plastic cup in the 50ml cylinder to measure the amount you collected. Record how much water you collected and which hole it came from.*

A 50ml cylinder will allow you to precisely measure the amount of dew you collected in each dew trap. Scientists frequently use graduated cylinders to measure the volume of liquids in experiments.

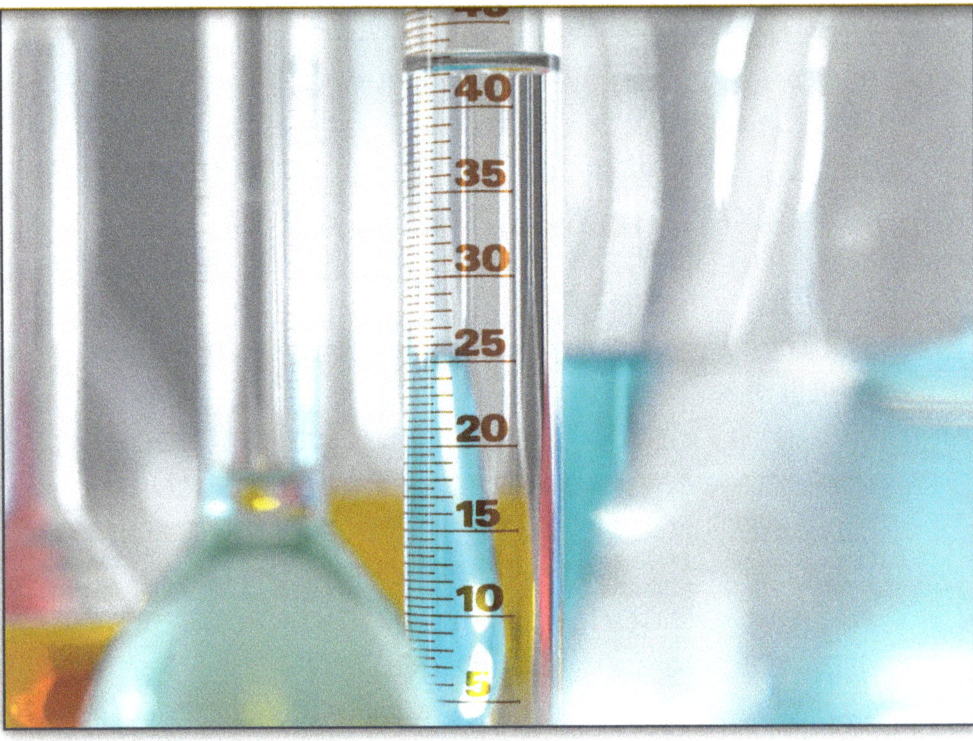

- *Repeat the collection, straining, and measuring process for the other holes.*
- *From your data, draw conclusions about how the diameter of the hole affects the amount of dew that can be collected. Prepare a presentation from the data you've collected.*

USEFUL TOOLS

These are just a few of the online tools that could prove useful for a science investigation that employs project-based learning. If you find other tools in your research, consider sharing the most helpful ones with your classmates. After all, science is a collaborative field!

- **Project-Based Learning:** Workbench, Dreamdo Schools
- **Concept-Mapping and Planning:** SimpleMind+, Evernote
- **Project Portfolios:** Pathbrite, Kidblog, Edublogs, Three Ring, WeLearnedIt, Seesaw
- **Making Presentations:** Prezi, PowerPoint, PresentationTube, Google Slides, Glogster
- **Adding Narration:** VoiceThread, FotoBabble, Audioboo
- **Creating Websites and Blogs:** Weebly, WordPress
- **Charts and Graphs:** Visme, Beam, ChartGo, NCES's Create a Graph
- **Science Project Ideas:** Science Buddies, Science Fair Central, Student Science, Education.com, Science-projects.org, Science Bob, sciencing.com

GLOSSARY

adaptation The process of changing to meet a changing environment or situation.

amendment In gardening, something added to the soil to improve its integrity.

atom The basic unit of a chemical element.

biodiversity The life in a particular habitat or ecosystem.

climate change Change in global or regional climate patterns.

drought-tolerant Able to withstand dry conditions.

global warming The gradual increase in the overall temperature of Earth's atmosphere.

hypothesis A proposed explanation based on limited evidence that is used as the starting point for future exploration.

inanimate Not alive.

kinetic energy Energy that a body possesses by being in motion.

magnitude In earthquake science, a numerical value assigned to rate the force of an earthquake.

Milky Way The galaxy in which the sun is located.

molecule A group of atoms bonded together.

mutation A change in the structure of a gene.

organic Derived from living matter.

prototype A preliminary model.

seismologist A scientist who studies earthquakes and related phenomena.

succulent In botany, a plant with thick, fleshy leaves or stems that are used to store water.

tectonic plates Large plates that make up Earth's crust.

trait A distinguishing quality or characteristic, especially one that is determined genetically.

velocity Speed.

verdant Green and lush.

FOR MORE INFORMATION

Canadian Wildlife Federation
350 Michael Cowpland Drive
Kanata, ON K2M 2W1
Canada
(800) 563-9453
Website: www.cwf-fcf.org/en
Twitter: @CWF_FCF
The Canadian Wildlife Federation works toward a sustainable, environmentally friendly future. It offers a wealth of information about Canada's wildlife, bodies of water, environments and ecosystems, and natural resources.

National Aeronautics and Space Administration (NASA)
300 E. Street SW, Suite 5R30
Washington, DC 20546
(202) 358-0001
Website: http://www.nasa.gov
Facebook and Twitter: @NASA
For all things space and solar system, NASA has been the United States' prime resource for decades.

National Wildlife Federation
PO Box 1583
Merrifield, VA 22116
(800) 822-9919
Website: http://www.nwf.org
Facebook: @NationalWildlife
Twitter: @NWF
The National Wildlife Federation is the oldest conservation organization in the United States. Its goal is to preserve US natural resources and habitats, ensuring a sustainable future for species that populate the continent.

Natural Resources Canada
(855) 525-9293
Website: www.nrcan.gc.ca/home
Twitter: @NRCan
This government agency is responsible for earth sciences, energy, resources, and environment in Canada.

United States Department of Energy
1000 Independence Avenue SW
Washington, DC 20585

(202) 586-5000
Website: http://www.energy.gov
This is the government agency that oversees energy usage and handling in the United States. Its programs include exploration and use of renewable energy sources.

United States Environmental Protection Agency
1200 Pennsylvania Avenue NW
Washington, DC 20460
Website: http://www.epa.gov
Facebook and Twitter: @EPA
This government agency is in charge of protecting the environment in the United States and, by extension, protecting human health.

United States Geological Survey (USGS)
12201 Sunrise Valley Drive
Reston, VA 20192
(703) 648-5953
Website: http://www.usgs.gov
Facebook: @USGeologicalSurvey
Twitter: @USGS
Part of the U.S. Department of the Interior, this government agency studies the natural resources in the United States as well as hazards. It is well known for its work on earthquakes, but it studies other environmental and natural hazards, too.

FOR FURTHER READING

Craig, Janet. *Safeguarding Water and Food Supplies.* New York, NY: Rosen Publishing, 2013.

Dickmann, Nancy. *Harnessing Wind Energy.* New York, NY: Rosen Publishing, 2017.

Engdahl, Sylvia. *Energy Alternatives.* New York, NY: Greenhaven Publishing, 2015.

Gerber, Larry. *Adapting to Droughts.* New York, NY: Rosen Publishing, 2013.

Gitlin, Martin. *Our Climate Future.* New York, NY: Greenhaven Publishing, 2018.

Hanson-Harding, Brian. *Composting: Urban Gardening and Farming for Teens.* New York, NY: Rosen Publishing, 2014.

Hawkins, Jay. *Earthquake Disasters.* New York, NY: Rosen Publishing, 2012.

Hayes, Amy. *Building Bridges and Roads.* New York, NY: Rosen Publishing, 2016.

Royson, Angela. *The Race to Survive Climate Change.* New York, NY: Rosen Publishing, 2015.

Shea, Therese M. *Rocked by Earthquakes.* New York, NY: Rosen Publishing, 2018.

BIBLIOGRAPHY

Climate Reality Project, "The Facts About Climate Change and Drought." June 15, 2016. https://www.climaterealityproject.org/blog/facts-about-climate-change-and-drought.

Duddu, Praveen. "Top 10 Biggest Wind Farms." Power Technology, September 29, 2013. https://www.power-technology.com/features/feature-biggest-wind-farms-in-the-world-texas.

Fong, Jen, and Paula Hewitt. "Worm Composting Basics." Cornell Composting. http://compost.css.cornell.edu/worms/basics.html.

Houghton, Colin. "Using Trees to Estimate Average Wind Speed: The Griggs-Putnam Index of Plant Deformity." My Wind Power System, March 26, 2012. https://www.mywindpowersystem.com/2012/03/26/using-trees-to-estimate-average-wind-speed-the-griggs-putnam-index-of-plant-deformity/#lightbox/2.

Kornei, Katherine. "Here Are Some of the World's Worst Cities for Air Quality." Science, March 21, 2017. http://www.sciencemag.org/news/2017/03/here-are-some-world-s-worst-cities-air-quality.

Lamb, Robert, and Michael Morrissey. "How Bridges Work." How Stuff Works. https://science.howstuffworks.com/engineering/civil/bridge3.htm.

Lau, Scott. "7 Weird Ways to Save Water." National Geographic Blog, August 5, 2014. https://blog.nationalgeographic.org/2014/08/05/7-weird-ways-to-save-water.

Modern Survivor Magazine. "Sourcing Water from Plants." http://survivor-magazine.com/tag/dew.

National Science and Technology Centre. "Oil and Water." https://www.questacon.edu.au/outreach/programs/science-circus/videos/oil-and-water.

Physics Classroom, The. "Newton's Third Law." http://www.physicsclassroom.com/Class/Newtlaws/U2L4a.cfm.

Socratic.org. "What Is Diffusion? What Is Facilitated Diffusion? What Is Osmosis?" May 22, 2017. https://socratic.org/questions/what-is-diffusion-what-is-facilitated-diffusion-what-is-osmosis.

Sohn, Emily. "The Power of Worm Poop." NPR, August 12, 2016. https://www.npr.org/sections/goatsandsoda/2016/08/12/489748719/the-power-of-worm-poop.

Tasteful Garden, The. "Worm Castings – What Are They? How Do They Work?" https://www.tastefulgarden.com/Worm-Castings-d114.htm.

Union of Concerned Scientists. "Causes of Drought: What's the Climate Connection?" https://www.ucsusa.org/global-warming/science-and-impacts/impacts/causes-of-drought-climate-change-connection.html#.WqlpMWbMyMI.

INDEX

ABOUT THE AUTHOR

Cathleen Small is a writer and editor. She has edited numerous books on technology and written multiple books for children and young adults. Growing up in technology-focused Silicon Valley, Small has always had a keen interest in science. When she's not writing or editing, she enjoys traveling and having adventures with her husband and two young sons.

PHOTO CREDITS

Cover, p. 1 Klaus Vedfelt/DigitalVision/Getty Images; p. 3 Ollyy /Shutterstock.com; p. 4 Pressmaster/Shutterstock.com; p. 5 Lapina /Shutterstock.com; p. 9 Pavel L Photo and Video/Shutterstock .com; p. 12 Science Photo Library/Getty Images; p. 13 FloridaStock /Shutterstock.com; p. 17 OlegDoroshin/Shutterstock.com; p. 20 Bruce C. Murray/Shutterstock.com; p. 22 Ron Bull/Alamy Stock Photo; p. 25 herjua/iStock/Thinkstock; p. 28 Alvin E Staffan/Science Source/Getty Images; p. 29 LisaInGlasses/E+/Getty Images;p. 32 U.S. Department of Energy; p. 35 Jaroslaw Grudzinski/Shutterstock .com; p. 38 Monkey Business Images/Shutterstock.com; p. 40 Eilis Maynard/FEMA; p. 41 Kamira/Shutterstock.com; p. 44 Filip Fuxa /Shutterstock.com; p. 46 Carlos Davila/Photographer's Choice RF /Getty Images; p. 48 David Greitzer/Shutterstock.com; p. 51 Alex Potemkin/E+/Getty Images;p. 54 Andrew Brookes/Cultura/Getty Images.

Design: Nelson Sá; Layout: Ellina Litmanovich; Editor: Amelie von Zumbusch; Photo Researcher: Nicole DiMella